Poh Chick loses the Moon

Axel Domínguez

Octavio Domínguez

Marielena Domínguez

Copyright © 2019 by Dr. Jorge Axel Domínguez-López, Octavio A. Domínguez-Durán and Marielena E. Domínguez-Durán.

All rights reserved. No part of this publication may be reproduced, distributed, or transmitted in any form or by any means, including photocopying, recording, or other electronic or mechanical methods, without the prior written permission of the publisher, except in the case of brief quotations embodied in critical reviews and certain other noncommercial uses permitted by copyright law.

Derechos Reservados © 2019 por Dr. Jorge Axel Domínguez-López, Octavio A. Domínguez-Durán and Marielena E. Domínguez-Durán.

Todos los derechos reservados. Ninguna parte de esta publicación puede ser reproducida, almacenada en un Sistema de recuperación de datos o transmitida de cualquier forma o por cualquier medio, electrónico, mecánico, fotocopiado, grabación o cualquier otro, sin la autorización por escrito del propietario de los derechos reservados y los autores.

ISBN: 978-1-708-70971-6

www.colibri.press

To our dear family.

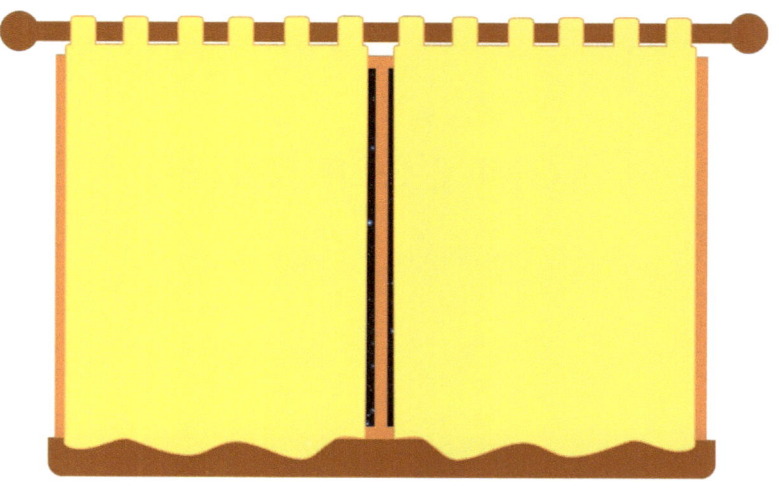

One night, Poh Chick could not fall asleep.

Poh Chick had a lot of energy and started jumping on his bed.

Hearing all the noise, Dad Rooster went to Poh Chick's room.

Why haven't you fallen asleep yet? - Dad Rooster asked as Poh Chick does a pirouette on the bed.

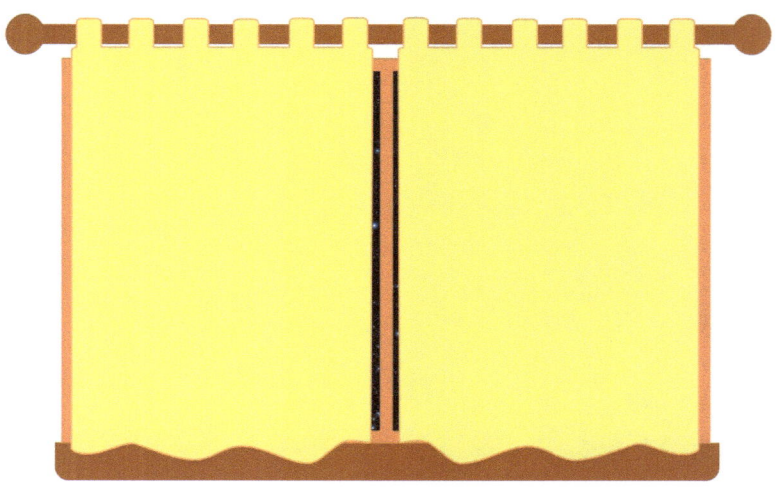

I'm not tired - Poh Chick replied. I really want to jump - he added.

I understand that you really want to keep playing but it's already time to sleep - said Dad Rooster.

Look, it's already night! - Dad Rooster exclaimed as he opened the curtains.

Poh Chick was surprised to see the darkness of the night. It was his first time seeing the sky at night. He went to the window to see in more detail and realized several things were white lights dotting the sky.

What are those tiny lights in the sky? - asked Chicky Poh.

Those lights are stars - answered Dad Rooster. The stars are like the Sun but because they are far away, they look small - added Dad Rooster.

Poh Chick observed that there was another light with a smile-like shape.

What is that funny light? - Poh Chick asked with intrigue.

That's the Moon - answered Dad Rooster while yawning.

Is the Moon a star? - Poh Chick quizzed.

The Moon is a satellite, a spheric rock that revolves around the Earth - explained Dad Rooster as he yawned again.

Thanks for the explanation, dad. Now I am going to sleep - said Poh Chick while giving him a hug and a good night kiss.

Poh Chick lay down on his bed and although many questions filled his head, he soon fell asleep.

Since that night, Poh Chick was fascinated with the Moon. Before falling asleep he looked out the window to see the Moon.

As the nights passed, Poh Chick realized two very interesting things: The Moon was not always in the same place and was changing shape. The first time he saw the Moon, it looked like a small smile. Now it looked like a big smile.

As the days went by, the Moon was no longer looking like a smile...

A few days later, like every night, Poh Chick looked out to see the Moon before falling asleep. He looked for her all over the sky but couldn't find her.

Dad! - Pollito Poh shouted at the top of his lungs.

Dad Rooster came running and very worried asked - What's up son? What is the emergency?

I have lost the Moon! - Poh Chick exclaimed while trying to contain the tears.

After calming down, Dad Rooster looked out of the window to help Chicky Chick search for the Moon. Indeed, there was no trace of the Moon that night.

Don't worry son, the Moon is not lost - said Dad Rooster while hugging Pollito Poh to calm him down. Right now we find out where she is - added Dad Rooster.

Yesterday before I fell asleep the Moon was still there - said Chicky Poh. What shape was the Moon last night? - asked Dad Rooster. Poh Chick grabbed paper and crayons, then began to draw. Look! The Moon was this shape last night! - exclaimed Chicky Poh.

When Dad Rooster saw the drawing, he knew immediately what had happened to the Moon that night. That form of the Moon is called the waning convex phase. It means that tonight the Moon will come out later than yesterday - explained Dad Rooster.

Tonight, you can sleep later so you can see the Moon - said Dad Rooster.

Thank you very much, Dad! - Poh Chick replied with a smile as big as the Moon.

A while later, the Moon appeared. Poh Chick jumped for joy at the same time he said - There she is! The Moon was not lost!

Indeed, the Moon was not lost. As the Moon revolves around the Earth and the Earth revolves itself, the Moon rises at different times - explained Dad Rooster to Chicky Poh.

There are days when the Moon rises during the day and not at night. If you wake up early tomorrow, you can see her again - added Dad Rooster. Poh Chick fell asleep soon.

The Sun had barely risen when Poh Chick woke up very excited to see the Moon.

Poh Chick continued to enjoy the Moon at night and day.

Moon Phases

www.ingramcontent.com/pod-product-compliance
Lightning Source LLC
Chambersburg PA
CBHW040253220526
45473CB00001B/461